GRAINS

by Robin Nelson

first step non-fiction

Lerner

Lerner Books · London · New York · Minneapolis

We need to eat many kinds
of food to be **healthy**.

We need to eat foods in
the **grains** group.

Grains are bread, cereal,
rice and pasta.

Grains give us **vitamins** and **minerals**.

Grains help food move
through our body.

Grains give us energy.

We need six **servings** of
grains each day.

We can eat bread.

We can eat waffles.

We can eat wraps.

We can eat cornflakes.

We can eat porridge.

We can eat rice.

We can eat spaghetti.

We can eat pretzels.

Grains keep me healthy.

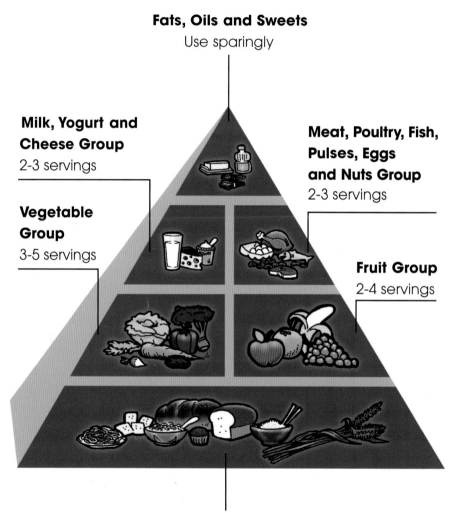

Fats, Oils and Sweets
Use sparingly

Milk, Yogurt and Cheese Group
2-3 servings

Meat, Poultry, Fish, Pulses, Eggs and Nuts Group
2-3 servings

Vegetable Group
3-5 servings

Fruit Group
2-4 servings

Bread, Cereal, Rice and Pasta Group
6-11 servings

Bread, Cereal, Rice and Pasta Group

The food pyramid shows us how many servings of different foods we should eat every day. The bread, cereal, rice and pasta group is on the bottom level of the food pyramid. The foods in this group are called grains. This part of the pyramid is the biggest because you need the most food from this group. You need to eat 6–11 servings of grains every day. Grains give you energy and make you strong.

Grains Facts

 Grains come from plants – mostly wheat, corn, rice and oats.

 Cereal, bread and pasta are made from wheat. A new kind of concrete is also made from wheat.

 White bread is good for you, but wholewheat bread is better for you. It has more fibre, which is very good for you.

There are many different kinds of pasta. You could try spaghetti, macaroni, tortellini, penne, ravioli, fettuccine or even bow tie.

 The average person in China consumes half a kilogram of rice a day.

 More foods are made from wheat than any other grain.

Glossary

 grains – seeds of wheat, corn, rice or oats

 healthy – not sick; well

 minerals – parts of food that keep your blood, bones and teeth healthy

 servings – amounts of food

 vitamins – parts of food that keep your body healthy

Index

This book was first published in the United States of America in 2003.

First published in the United Kingdom in 2008 by
Lerner Books,
Dalton House,
60 Windsor Avenue,
London SW19 2RR

Website address: www.lernerbooks.co.uk

This edition was updated and edited for UK publication by Discovery Books Ltd., Unit 3, 37 Watling Street, Leintwardine, Shropshire SY7 0LW

Words in **bold** type are explained in the glossary on page 22.

British Library Cataloguing in Publication Data

Nelson, Robin, 1971-
  Grains. - (First step non-fiction. Food groups)
  1. Grain in human nutrition - Juvenile literature 2. Grain
  - Juvenile literature
  I. Title
  641.3'31

  ISBN-13: 978 1 58013 390 6

The photographs in this book are reproduced through the courtesy of: © Todd Strand/Independent picture Service, front cover, pp 4, 12, 13, 14, 17; © PhotoDisc/Getty Images, pp 2, 5, 7, 9, 10, 22 (second from top, middle, bottom); © Wheat Foods Council, pp 3, 8, 11, 15, 16, 22 (top, second from bottom); © Royalty-Free/CORBIS, p 6.

The illustration on page 18 is by Bill Hauser/Independent Picture Service.

Printed in China